The World's Oceans

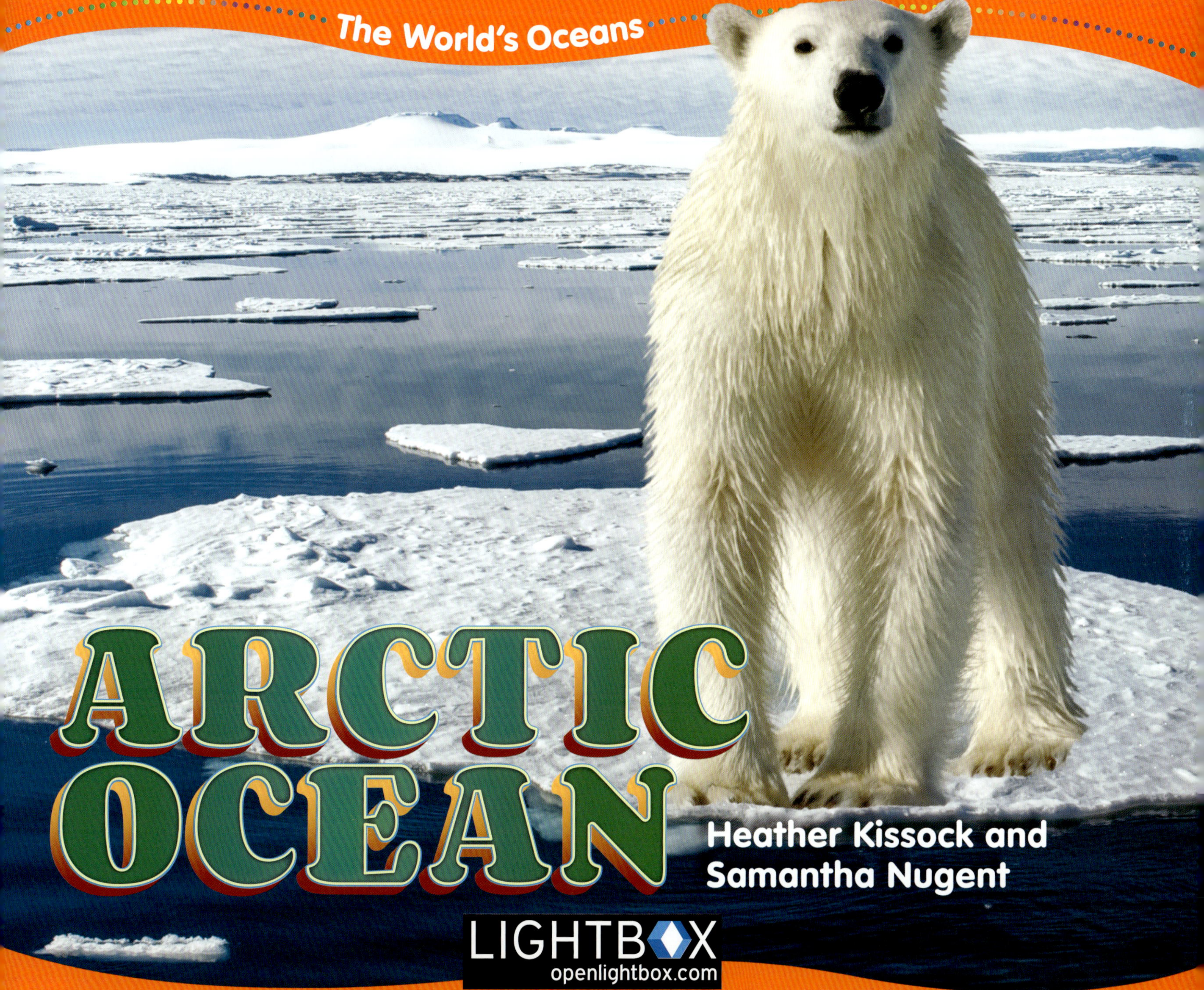

ARCTIC OCEAN

Heather Kissock and
Samantha Nugent

LIGHTBOX
openlightbox.com

LIGHTBOX

Go to
www.openlightbox.com
and enter this book's
unique code.

ACCESS CODE

LBXN4793

Lightbox is an all-inclusive digital solution for the teaching and learning of curriculum topics in an original, groundbreaking way. Lightbox is based on National Curriculum Standards.

OPTIMIZED FOR

- ✔ TABLETS
- ✔ WHITEBOARDS
- ✔ COMPUTERS
- ✔ AND MUCH MORE!

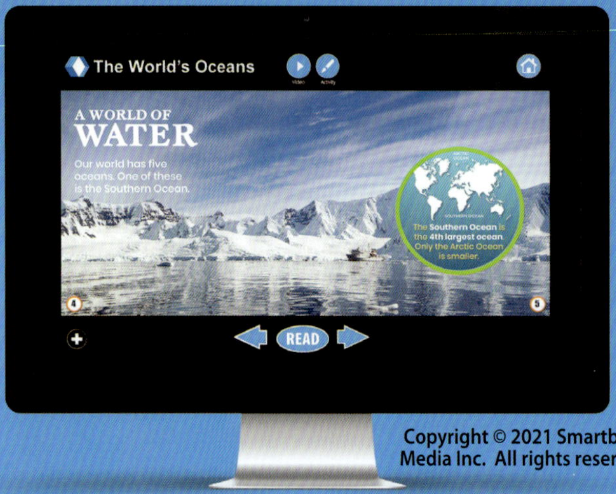

STANDARD FEATURES OF LIGHTBOX

AUDIO High-quality narration using text-to-speech system

VIDEOS Embedded high-definition video clips

ACTIVITIES Printable PDFs that can be emailed and graded

WEBLINKS Curated links to external, child-safe resources

SLIDESHOWS Pictorial overviews of key concepts

INTERACTIVE MAPS Interactive maps and aerial satellite imagery

QUIZZES Ten multiple choice questions that are automatically graded and emailed for teacher assessment

KEY WORDS Matching key concepts to their definitions

VIDEOS

WEBLINKS

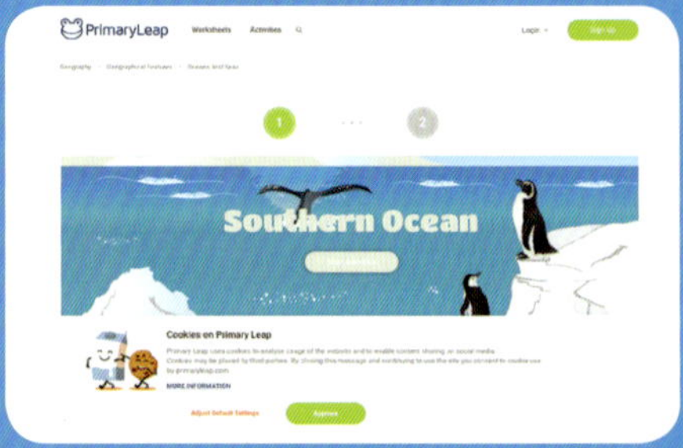

SLIDESHOWS

QUIZZES

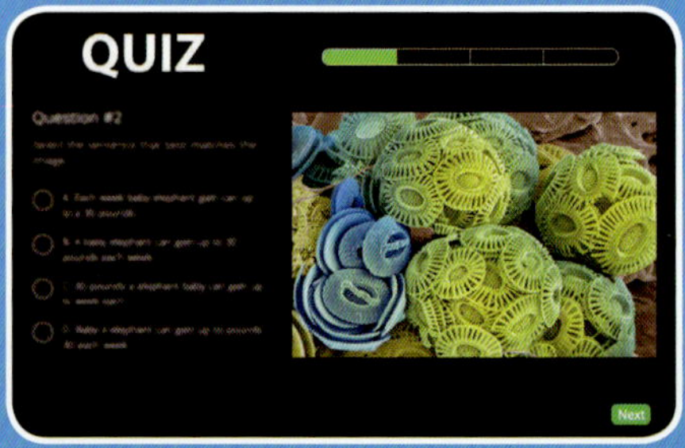

The World's Oceans

ARCTIC OCEAN

Contents

A WORLD OF WATER

Earth is home to five oceans. The Arctic Ocean is Earth's smallest ocean.

The Arctic Ocean is about **1.5 times** as big as the **United States.**

WHERE IS THE ARCTIC OCEAN?

The Arctic Ocean surrounds the North Pole. Its waters link with two other oceans.

THE WORLD'S OCEANS

ARCTIC OCEAN FEATURES

Ice covers the Arctic Ocean for much of the year. More than half of its ice can melt during the summer.

Scientists predict that the Arctic Ocean will have **ice-free summers** within the **next 20 years.**

ANIMALS OF THE ARCTIC OCEAN

Many different animals live in and around the Arctic Ocean. Some are found nowhere else on Earth.

Polar Bear

Arctic Cod

Ivory Gull

Ringed Seal

Copepod

PLANTS OF THE ARCTIC OCEAN

The cold waters of the Arctic Ocean are home to several unique plants. They provide food for the animals that live there.

Ice Algae

Phytoplankton

Cuvie Kelp

ARCTIC ECOSYSTEM

The plants and animals that live in and around the Arctic Ocean need each other to survive. Together, they form an ecosystem.

ARCTIC FOOD CHAIN

Ice algae grows under the sea ice. It is food for copepods.

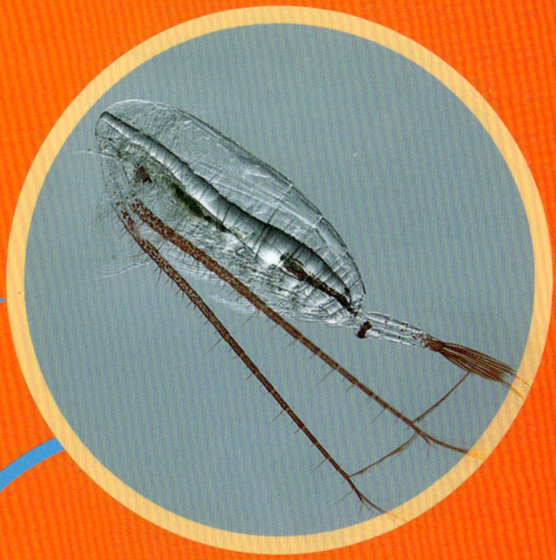

Copepods are food for Arctic cod.

Arctic cod are one kind of fish eaten by ringed seals.

Ringed seals are a polar bear's main food.

OCEAN IN DANGER

Companies drill for oil and gas in the Arctic Ocean.

Oil spills can harm the ocean's waters and the animals that live in them.

LEARNING MORE

Some people study the Arctic Ocean.

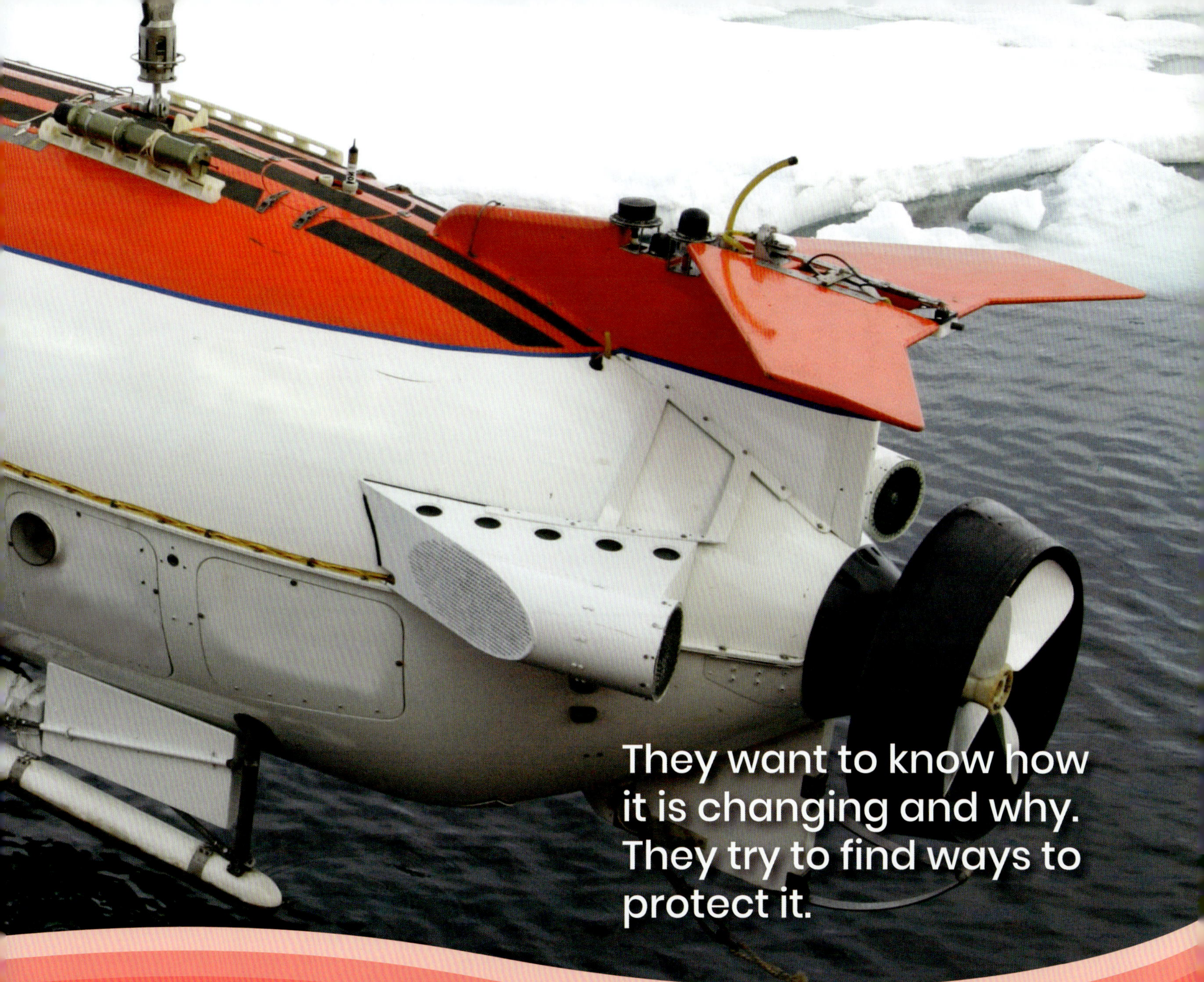

They want to know how it is changing and why. They try to find ways to protect it.

HELPING THE OCEAN

You can help the Arctic Ocean.

Turn off lights when you are not using them.

Many homes get their electricity from oil and gas.

ARCTIC OCEAN FACTS

This page provides more detail about the interesting facts found in the book. Simply look at the corresponding page number to match the fact.

Pages 4–5

The Arctic Ocean covers an area of about 5,440,000 square miles (14,090,000 square kilometers). The United States, on the other hand, covers about 3.7 million square miles (9.5 million sq. km). The Arctic Ocean is also the coldest of the world's oceans. Its temperature ranges from about 37° Fahrenheit (3° Celsius) to 32°F (0°C).

Pages 6–7

The Arctic Ocean touches the borders of Canada, Greenland, Iceland, Russia, Norway, and the state of Alaska. South of the Arctic Ocean are the Pacific Ocean and the Atlantic Ocean. The Arctic Ocean itself includes eight seas.

Pages 8–9

Global climate change has caused sea ice on the Arctic Ocean to shrink and water levels to rise. With less sea ice in the way, wind creates stronger waves on the Arctic Ocean. Warmer temperatures have also caused the Arctic Ocean's coastline to soften as frozen soil thaws. Stronger waves and softer soil have resulted in the rapid erosion of northern coastlines.

Pages 10–11

The Arctic Ocean is home to many types of animals. Several have adaptations that make them well suited to the ocean's cold waters. Polar bears and seals rely on blubber to keep them warm. Arctic cod have proteins that ensure they do not freeze in the cold waters.

Pages 12–13

Much of the ocean's plant life is found near the water's surface. Ice algae tends to attach itself to the underside of sea ice. Some, such as the *Melosira arctica*, can form large mats on the ice. Phytoplankton lives near the surface, where it has access to sunlight.

Pages 14–15

All living things have survival needs, such as food, water, and habitats that allow them to grow. A community of interconnected organisms is called an ecosystem. Producers such as plants create their own food from the Sun's energy. Consumers can then eat producers or other consumers to gain energy. This cycling of energy is called a food chain.

Pages 16–17

The world's largest untapped supply of natural gas lies beneath Arctic waters. Accessing it would help heat homes and run factories for years to come. Drilling introduces a number of dangers. The noise produced from drilling has been proven to change animal behavior. Oil spills contaminate the water and kill wildlife.

Pages 18–19

As climate change makes the Arctic Ocean's waters increasingly ice free, people can access the area with greater ease. More human activity creates significant issues for the ocean, as well as the plants, animals, and people that rely on it. Scientists from organizations ranging from NASA to Greenpeace are all actively engaged in finding ways to protect this ocean from further harm.

Pages 20–21

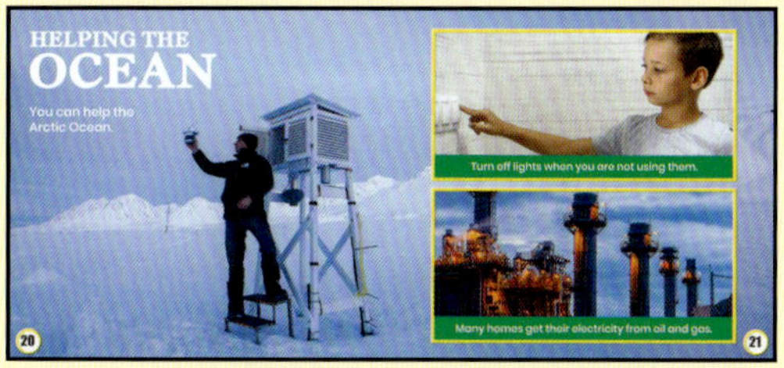

When burned to make energy, oil and gas send carbon dioxide into the atmosphere. This carbon dioxide traps heat near Earth's surface, making it warmer than it should be. This causes climate change. If people were to reduce their use of oil and gas, climate change could be slowed. If the need for oil and gas was reduced, people might not need to drill for it in the Arctic.

KEY WORDS

Research has shown that as much as 65 percent of all written material published in English is made up of 300 words. These 300 words cannot be taught using pictures or learned by sounding them out. They must be recognized by sight. This book contains 75 common sight words to help young readers improve their reading fluency and comprehension. This book also teaches young readers several important content words, such as proper nouns. These words are paired with pictures to aid in learning and improve understanding.

Page	Sight Words First Appearance
4	a, Earth, home, is, of, the, to, water, world
5	about, as, big, states, times
6	its, other, two, where, with
8	can, for, more, much, than, year
9	have, next, that, will
10	and, animals, are, around, different, found, in, live, many, on, some
12	food, plants, there, they
14	an, each, need, together
15	by, grows, it, kind, one, sea, under
17	them
18	people, study
19	find, how, know, try, want, ways, why
20	help, you
21	from, get, lights, not, off, their, turn, when

Page	Content Words First Appearance
4	Arctic Ocean, oceans
5	United States
6	North Pole
7	Africa, Antarctica, Asia, Atlantic Ocean, Australia, Europe, Indian Ocean, North America, Pacific Ocean, South America, Southern Ocean
8	features, half, ice, summer
9	scientists
11	Arctic cod, copepod, ivory gull, polar bear, ringed seal
13	cuvie kelp, ice algae, phytoplankton
14	ecosystem
15	fish, food chain
16	companies, danger, gas, oil
17	spills
21	electricity

Published by Smartbook Media Inc.
14 Penn Plaza 9th Floor
New York, NY 10122
Website: www.openlightbox.com

Library of Congress Cataloging-in-Publication Data
Names: Kissock, Heather, author. | Nugent, Samantha, author.
Title: Arctic Ocean / Heather Kissock and Samantha Nugent.
Description: New York : Smartbook Media, Inc., 2021. | Series: The world's oceans | Audience: Ages 6-8 | Audience: Grades 2-3 |
Identifiers: LCCN 2020018603 (print) | LCCN 2020018604 (ebook) | ISBN 9781510555167 (library binding) | ISBN 9781510555174
Subjects: LCSH: Arctic Ocean--Juvenile literature. | Oceanography--Arctic Ocean--Juvenile literature. | Marine ecology--Arctic Ocean--Juvenile literature.
Classification: LCC GC401 .K57 2021 (print) | LCC GC401 (ebook) | DDC 551.46/132--dc23
LC record available at https://lccn.loc.gov/2020018603
LC ebook record available at https://lccn.loc.gov/2020018604

052020
110819

Printed in Guangzhou, China
1 2 3 4 5 6 7 8 9 0 24 23 22 21 20

Editor: Heather Kissock Designer: Jean Rodriguez

Every reasonable effort has been made to trace ownership and to obtain permission to reprint copyright material. The publisher would be pleased to have any errors or omissions brought to its attention so that they may be corrected in subsequent printings.
The publisher acknowledges Getty Images, Shutterstock, and Wikimedia as its primary image suppliers for this title.